Raspberry Pi 2

The Ultimate Beginner's Guide!

Andrew Johansen

© **Copyright 2015 by Andrew Johansen - All rights reserved.**

This document is geared towards providing exact and reliable information in regards to the topic and issue covered. The publication is sold with the idea that the publisher is not required to render accounting, officially permitted, or otherwise, qualified services. If advice is necessary, legal or professional, a practiced individual in the profession should be ordered.

- From a Declaration of Principles which was accepted and approved equally by a Committee of the American Bar Association and a Committee of Publishers and Associations.

In no way is it legal to reproduce, duplicate, or transmit any part of this document in either electronic means or in printed format. Recording of this publication is strictly prohibited and any storage of this document is not allowed unless with written permission from the publisher. All rights reserved.

The information provided herein is stated to be truthful and consistent, in that any liability, in terms of inattention or otherwise, by any usage or abuse of any policies, processes, or directions contained within is the solitary and utter responsibility of the recipient reader. Under no circumstances will any legal responsibility or blame be held against the publisher for any reparation, damages, or monetary loss due to the information herein, either directly or indirectly.

Respective authors own all copyrights not held by the publisher.

The information herein is offered for informational purposes solely, and is universal as so. The presentation of the information is without contract or any type of guarantee assurance.

The trademarks that are used are without any consent, and the publication of the trademark is without permission or backing by the trademark owner. All trademarks and brands within this book are for clarifying purposes only and are the owned by the owners themselves, not affiliated with this document.

Table of Contents

Introduction v

Chapter 1: Overview of Raspberry Pi 2 1

Chapter 2: Getting Familiar with the Hardware 5

Chapter 3: The Operating System 25

Chapter 4: Integrating Raspberry Pi 2 31

Chapter 5: Linux 39

Chapter 6: Raspberry Management 51

Chapter 7: Using Raspberry 55

Chapter 8: Raspberry Pi: More than a personal computer 59

Conclusion 63

Introduction

I want to thank you and congratulate you for getting my book…

"Raspberry Pi 2 – The Ultimate Beginner's Guide"

For many years now people have been content with merely knowing how to perform repetitive manual functions on their computers. Just like any household device, our relationship with the computer is limited to purchasing it, unpacking it from its box, adding the necessary software to make it function and then we start using it. For most people that's all the computer is about - a personal technology device that we use for specific purposes. No different from say - a toaster.

Why should it be any different? After all what purpose does the computer serve? It was invented as an electronic and technology device to allow people to work in a more efficient way. A gadget developed to give people the ability to optimize work and all other human activity its application is relevant to.

Since its invention computers have earned the reputation of being too technically complicated. It inner workings were just beyond the grasp of ordinary people. It was not necessary to understand the internal functions of the computer especially if it was working to specifications. To do so is just too intimidating.

This defeatist thinking created a wall that stumped any desire a person may have about knowing more about computers. It limited people's understanding of computers to the use of its built in programs. Shunning any inclination to get in-depth know-how of its functions and how computers are built.

Its seeming complicated nature has turned off people from developing an interest in understanding how it all worked much less want to learn how to build it themselves.

Since the device was invented, the computer has had this invisible sign saying, 'Hands Off. No Tinkering.' For years computers, have been considered the avenue of engineers and scientists, as if it was something beyond the grasp of ordinary people. Because of this, there is a lack of motivation for executing complicated tasks that can lead to new program creation. We have limited computer education to knowing how programs worked and not so much how the actual computer hardware worked.

Often computer courses are dedicated to how-to lessons on software utility with no regard to understanding hardware functionality. People have been limited to understanding how the computer thinks and process information. We have never been encouraged to analyze how the computer was assembled to accomplish its function.

Now this is where Raspberry Pi comes in. Learning how software and hardware work hand-in-hand to achieve a desired result is the goal of Raspberry Pi. It gives people the ability to develop not only computer programs but also invent actual computer & electronic based devices.

This book will provide you with basic computing knowledge that will allow you to invent any device that can be powered by the RaspBerry Pi. Your imagination is the only limitation. So get creative. Who knows, you might just be the next big name in Silicon Valley!

Thanks again for purchasing this book, I hope you enjoy it!

CHAPTER 1

Overview of Raspberry Pi 2

Raspberry Pi can be used for word and data processing, listening to music, watching videos, playing games and photo editing just like any ordinary computer but it has a lot more to offer than just that. It bridges a world where electronics and programming can be a simple Do-It-Yourself project.

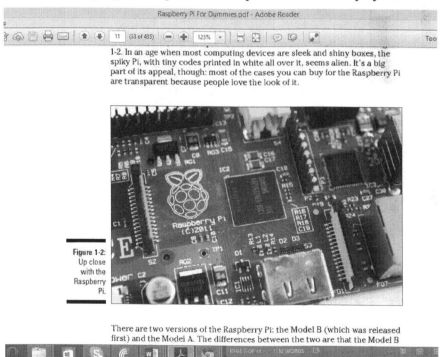

Figure 1: Raspberry Pi 1

For example, recently it's been taken up by none other than the astronauts in the International Space Station. Tim Peake is a British astronaut who is on a mission in the ISS. He put forward a Raspberry Pi challenge to student inventors; he asked them to create for him a Raspberry Pi powered MP3 player.

Peake loves listening to music when he works out. Since he trains daily on the International Space Station, he requires an MP3 player. He is not able to use his Apple iPod because it is not compatible with the ISS systems. The challenge he set for UK students was to build a Raspberry Pi MP3 player that is compatible with the ISS system that he can use to listen to music while in the ISS.

UK students studying Raspberry Pi technology took on the challenge and soon enough the winning entry was launched into space and received by Tim Peake in the International Space Station. Now the British astronaut is able to listen to his favorite tunes while exercising on the ISS all thanks to a group of young students who could code using Raspberry Pi technology.

That is just one of many stories about the Raspberry Pi. There is a lot to thank for this new piece of technology. Raspberry Pi opens up several doorways of opportunities for everyone. It may look like a small credit card-sized circuit board but it allows you to create an infinitesimal number of projects.

A look back at History

The Raspberry Pi is a 3.5 x 2.5 inch rectangular single-circuit board computer. Let's compare it to the size of a credit card. It was developed in England, United Kingdom by the Raspberry Foundation. Initially it was invented to aid computer science courses in schools. The people at Raspberry Foundation want to encourage students to develop an interest in coding by giving them the ability to create their own computers from scratch.

It was simple enough to enable the teaching of computer programming to kids. Since its launch, the Raspberry program has received very favorable response. As of 2015, it has sold six million units. As of this moment, it is considered England's fastest selling personal computer and one of the world's most popular versions of the computer circuit board.

The ultimate goal of Raspberry Foundation for developing Raspberry Pi is to enable the teaching of computer science in as early as grade school. They want to encourage coding in the next generation of the workforce. And the focus is not just in the UK but all around the world with emphasis on developing countries.

For most people, computers are an intimidating piece of technology. With complicated electronic circuitry and even more complicated software programming. For the average person, the computer is a cryptic device filled with codes and wires only the technologically savvy computer programmer can understand.

The Raspberry Pi Foundation aims to shatter that perception about computers. They want people to realize that computers are just like any other technological or electronic device. They want people everywhere in the world to know it is merely a combination of technological hardware and software whose assembly is as easy to do as any electronic gadget.

Raspberry Pi believes if people found the time to deconstruct a computer and disassemble it part by part, they will get to see every hardware or part that make up the device and understand how they function. If people do this, they will realize that computers are not as complicated a device as they have lead themselves to believe. For the Raspberry Pi Foundation if people are willing to understand how computers are built and made operational they have it in themselves to create their own version of a personal computer.

With the Raspberry Pi, their options are not limited to merely creating their own personal computer – they can create virtually any electronic device they can imagine! After all, the Raspberry Pi is a circuit board that can automate anything.

For example, it can power MP3 players. Audiophiles can develop their own MP3 players according to their preferred specifications. If you're interested in security systems, you can also create laser-operated alarm systems. The Raspberry Pi can automate sound or voice sensitive lighting systems for any space. It can be used to create infrared cameras to monitor nocturnal activities of animals. The possibilities for inventing new devices are endless with the Raspberry Pi. If only people would take the time to understand how it worked.

For the Raspberry Pi Foundation, to revolutionize the world we have to start teaching kids to get comfortable with the language of computer science and computer programming. We have to create a new breed of people who build computers and not just use them. There is potential to create a better world with the Raspberry Pi.

With it kids get to understand the inner workings of computers so that they can build it themselves. It gives them the knowledge and tools to build their own version of a computer from scratch. With Raspberry Pi technology kids will be able to get comfortable with computer hardware. They will learn how to piece together each piece of computer hardware and circuitry and power it so that it is a fully functional computing device.

Working with Raspberry Pi shatters the illusion that computer science and computer programming is just the milieu of grown-ups and engineers or computer science experts. Raspberry Pi makes building a computer as easy as riding a bike. Kids will realize that once they get the basics, the sky is the limit to the variety of inventions or new gadgets that they will be able to create. They will see that this credit-card side piece of circuitry can make inventors of them.

Since its launch, Raspberry Pi has earned the interest of not just the education sector but also of national governments everywhere in the world. The Raspberry Foundation vision has caught on. Leaders have understood that the Raspberry Pi has the potential to reshape a country's future. It is the sort of technology that can be a game-changer. It has the ability to equip the next generation of a country's workforce with the sophisticated knowledge and skills of computer science and programming. Raspberry Pi can exponentially grow the intellectual and technological skills of a country's workforce by making them tech savvy enough to be able to build their own computers and tech gadgets.

CHAPTER 2

Getting Familiar with the Hardware

When you get a hold of a Raspberry Pi – regardless if it's the first release or the second release – you will see that it is a circuit board with various sockets and components attached to it. Despite the trend to favor thin, sleek and less-detailed gadget design, the look of Raspberry Pi appeals to a lot of people.

The newer version of Raspberry Pi comes in two models: Model B and Model A. Model B was the first release; it has two USB ports, and memory up to 512 MB in its latest releases. Model A only has a single USB port and up to 256MB. Model B is sold for $35 whereas Model A is sold for $25. It has since been upgraded to a Model A+ and a Model B+ where additional USB ports have been added as well as additional memories.

Below is a diagram of the features of a Raspberry Pi circuit board:

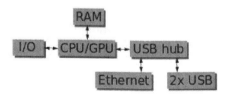

Depending on the model whether Raspberry Pi 1 or 2 model A, A+, B, B+, or Zero all Raspberry Pi models will have a USB and Ethernet port, containing a computer processing unit (CPU), a graphics processing unit (GPU), an I/O, and RAM. It is a simple design and believe it or not this is exactly what circuit boards look like in every computer, even the most sophisticated versions.

Raspberry was developed in the same time that mobile phone companies started manufacturing smaller chips in a race to come up with the thinnest hardware to accompany the best software. The people behind the project surmised that instead of signing up for the race, they can change the terrain altogether. It should be expected therefore that the system on the chip (SoC) in the Raspberry Pi are similar to the ones found in iPhone/3gs/3g smartphones.

The SoC or the heart and soul of Raspberry is a chip processor called Broadcom BCM2835 SoC. It includes a 700 MHzARM1176JZF-S processor or central processing unit (CPU). It also has a graphics processing unit (GPU) called Videocore IV (GPU). And of course a RAM because a computer circuit board cannot be complete without it.

Its RAM has 16KB level 1 cache and a 128KB level 2 cache for the GPU, the same for the CPU. Since the SoC of the Raspberry has both CPU and GPU it is able to deliver high-definition video and pictures. Without the combination CPU and GPU circuitry, it is not likely the Raspberry Pi device will be able to process 1080p high definition videos and pictures that the Raspberry Pi camera captures; the same with video and picture preview.

We have said earlier that the Raspberry Pi is similar to other circuit boards found inside commercially produced computers. For example, the CPU of the Raspberry Pi is equal to that of the CPU of the 1999 Pentium II while the GPU is equal in performance to Xbox 2001. It is expected to improve in the future as more upgraded versions are released. For now, for a circuit board students can use to tinker about with building computers and tech gadgets it serves its purpose.

RAM

All the Raspberry Pi models A, A+, B, B+ and Zero have 256MB RAM. The CPU gets 128MB RAM while the GPU gets the other 128MB RAM. That is more than sufficient memory for any device to function at an efficient speed. And if it has a camera function, to record and view high definition videos as well as preview or develop 3D animation.

It's good to note that some of the latest versions of smartphones and tablets have the same RAM configuration as the Raspberry Pi. As of early this year, the Raspberry Pi 2 has been recorded to have a 1GB RAM while the Raspberry Pi Zero has 512MB of RAM enabling both to process information at a higher speed and more efficient rate than previous versions.

These are configurations that will satisfy any inventor. Processing speed and ability to preview high-definition video and photos are minimum requirements these days for developing quality computers and gadgets.

Networking

If connection to an internet network is required it is better to use the Raspberry Pi model B and B+ since it has a built in USB Ethernet adapter. Not the same with the Raspberry Pi model A, A+, and Zero. Their Ethernet ports are external and not built in to the circuit board. The external USB Ethernet adapter on the A, A+ and Zero models can be connected to a USB hub.

Model Specifications

Raspberry Pi 1 and 2 Models A, A+, B, and B+ have almost similar features and configurations. All have a SoC of Broadcom BCM2835 which houses a CPU, GPU, DSP, SDRAM, and USB Port. The number of USB ports increase as the models get upgrades. For example the Raspberry 2 Model B and B+ have 2 USB ports.

All models have CPUs powered by a 700 MHz single-core ARM1176JZF-S. All GPUs are 250MHz BroadcomVideoCore with OpenGLES 2.0, MPEG & VC-1 licensed, and a 1080p/H.264/MPEG4- AVC high profile decoder and encoder.

The RAMs vary from 256MB to 512MB and a high of 1GB. All models have onboard video input and output with onboard storage and power source. Sizes are the same at 3.5 x 2.5 inches and all units weigh 4.5 grams.

Raspberry is run by Linux, an open source operating system. Not only does Linux have a different take on the technology industry; instead of protecting the OS from being copied by competitor, the people behind Linux enlists expert minds of volunteers who decided to work together. Their platform allows closer inspection with the ability to alter the source code to fit the needs of whatever program people have in mind.

Unless you have Linux installed, you won't be able to open other software in Raspberry Pi. Fortunately most of the software associated with Linux is free.

Additional Product Upgrades

Over the years, there have been additional product upgrades on the Raspberry Pi circuit board. Here are some of the newer versions:

The RaspBerry Pi 2 Model B

This second generation RaspBerry Pi replaced the original Raspberry Pi 1 Model B+. It was launched in February 2015. Its upgrade included an A 900MHz quad core ARM Cortex-A7 CPU and 1GB RAM. It kept most of the features of the original version. The Raspberry Pi 2 Model B also has 4 USB ports, 40 GPIO pins, Full HDMI port, camera interface, combined 3.5mm audio jack and composite video, display interface, micro SD card slot, and Videocore IV 3D graphics core.

With its ARMv7 processor, it can operate ARM GNU/Linux, distributions plus Snappy Ubuntu Core and Microsoft Windows 10. The Raspberry Pi 2 is compatible to Raspberry Pi 1. It is the perfect version for use in schools as it offers great flexibility for learners something which cannot be said about the other versions such as Model A+. It is the preferred model of educators and teachers.

The RaspBerry Pi 2 Model B+

It is considered the final version of Raspberry Pi. The Raspberry Pi 2 Model B+ replaced in July of 2014 the Model B. It differs from the Model B version in terms of:

Increased GPIO: Now with a GPIO (general purpose input and output) header containing 40 pins, it has the same number of pins for the 1st 26 pins as the Model A & B.

Increased USB: In the previous models there were only 2 USB ports. In this version now it has been increased to four with upgraded overcurrent and hot plug behavior.

New Micro SD card: No more friction-fit SD card socket. This version has a push-push micro SD.

Less power consumption: This version lowers power consumption between 0.5W and 1W. The linear regulators have been removed in favor of the switching version. It is even more quiet now when operational because of this power supply upgrade.

Improved audio: With the low noise power supply the audio circuit is so much better.

Upgraded form factor: This version has additional 4 square mounting holes, better aligned USB connectors with the edge of the board and composite video's been moved to the 3.5mm jack.

Much like the Model B, this version is equally user-friendly. Kids and young students the Raspberry Pi was developed for will be very comfortable working with this version.

The RaspBerry Pi Model A+

Considered the low cost version of the RaspBerry Pi, it is the replacement of the original Model A. The Model A was launched in November 2014. Its upgrades include additional GPIO with 40 more GPIO pins. It has push-push micro SD which is a better version of the previous friction-fit SD card.

Much like the Model B+ version, the Model A+ also has lower power consumption as it replaced the linear regulators with the switch version. Audio is so much more improved and it has a more streamlined form factor. This version is 2cm shorter than the Model A.

The Model A+ is for more advanced embedded projects especially for those that require low power, do not require multiple USB ports and Ethernet.

The RaspBerry Pi Zero

At $5 per unit it is the most affordable version of the RaspBerry Pi making it the most accessible RaspBerry Pi for all sorts of projects. It is also the smallest version of the RaspBerry Pi units. But do not let its size discriminate against its functionality. It has twice the utility of the Model A+.

The RaspBerry Pi Zero includes 1Ghz, single core CPU, 512MB RAM, Micro USB power, HAT-compatible 40-pin header, mini HDMI and USB ports, and composite video and reset headers.

Various Applications

When you turn on the Raspberry Pi, the first thing you'll notice is it's text prompt however you can connect it to a graphical desktop that streamlines program and application management. Aside from the functions enumerated in the previous section, Raspberry Pi allows you to browse the internet and even build a website. Almost everything that you can do with a computer, you can do with Raspberry Pi.

If you are a computer student, Raspberry Pi does not only bridge the knowledge gap between what the computer thinks and how it does what it thinks. This way, you learn the key concepts of hardware structures as it executes the instruction provided by the software.

Limitations

For a miniature chip that is inexpensive, Raspberry Pi is a powerful tool but it had limitations too. While the possibilities seemed endless for the chip, its power is similar to that of a mobile device than a high-powered PC.

Raspberry Pi Foundation described the first release is comparable with a 300 MHz Pentium 2-run PC, which was popular back during the late nineties. The only difference is that Raspberry Pi has much better graphics similar to Xbox gaming console released a decade ago.

The newest release – Raspberry Pi 2 – rectified this lag behind the circuit board addressing concerns such as memory limitation (from 256 MB to 512 MB), and processing capabilities.

While the system that runs both the Xbox and Pentium 2 PC were well-received q few years after they released, both systems can't compete now with the faster and sleeker consoles and processors. Thus, Raspberry Pi might not be able to forge its way ahead of the competition as it is unable to address the demand specifications of modern programs and software. Fortunately, it is easy to look for programs, install them, try them and delete them if it does not work well with Pi. Good news is that there are a lot of programs available that work well on Raspberry Pi.

Pi won't take the place of your main computer panel but it will give you enough flexibility for you to harness your creativity to do projects that you may not have dared to try if you were stuck with your more expensive PC.

TECHNICAL GLITCHES

It is necessary to note that the Raspberry Pi 2 Model B is prone to malfunction due to flashes of light. It has been observed that xenon camera flashes and laser pointers can cause the Raspberry Pi 2 to reboot and turn off automatically. If a camera is flashed and the laser pointer is directed towards the Raspberry Pi 2 chip it will cause it to malfunction. Other sources of light, such as light bulbs which are uninterrupted and continuous light sources, do not have the same effect on the chips.

Several tests were conducted to find out what was really causing the glitch. They first ruled out the possibility that the electromagnetic pulse from the camera flash was the cause. First they covered the chip while flashing the light. Then they flashed the light at the back of the circuit board. Both instance saw no malfunction. This proved EMP was out of the question as reason for malfunction.

To prove that flashing light was the cause of the malfunction, laser pointers were brought in. The Raspberry Pi 2 was tested for both camera flash and laser pointers. True enough the circuitry faltered. It was discovered that unless covered with a film of silicon the sudden flashes of light caused the output voltage of the power supply to falter triggering a reboot or total shut down of the system.

It is recommended that when working with a Raspberry Pi 2 Model B do not take photos of the circuit board or inspect it with laser pointers to ensure that it remains functional.

Understanding the hardware inside the Raspberry Pi

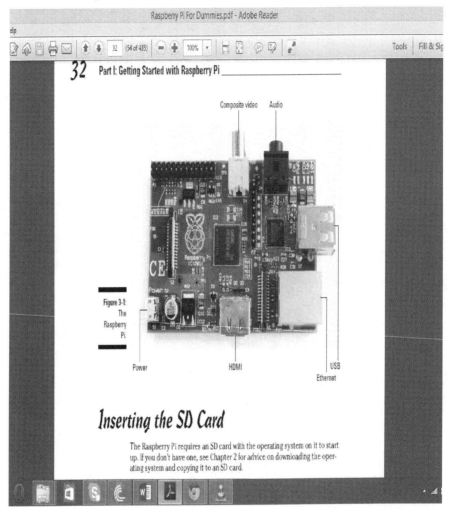

The above image details the hardware and components that make up the Raspberry Pi. Let us go through each feature in detail for a better understanding of how this particular circuit board functions.

System on the Chip (SoC): This is the integrated circuit that brings together all components of a computer in a single chip. For the Raspberry Pi it is where the CPU, GPU, DSP, and SDRAM are located. The SoC is the preferred system for mobile electronics such as smartphones because of their have low power consumption.

Central Processing Unit (CPU): the part of the computer that processes the instructions of computer programs. It is the electronic circuitry that performs the basic arithmetic, control, logical, and input/output functions based on the instructions or programming. It is the brain center of the Raspberry Pi.

Graphic Processing Unit (GPU): Also called the visual processing unit (VPU). It is an electronic circuit that can alter or manipulate memory to fast track creation of images so it can be displayed by the computer. GPUs are present in mobile phones, computers, game consoles, and workstations. Any tech gadget really that features picture display.

High-Definition Multimedia Interface (HDMI): the audio/video interface where video and audio data can be transferred from a digital video/audio device to the Raspberry Pi computer. The docket where cameras, computers, mp3 players, or projectors are plugged into for the purpose of transferring video and still images to the Raspberry Pi device.

Ethernet: the hardware in the Raspberry Pi that enables the gadget to connect to an internet network.

Composite Video: Analog video only transmission for standard definition video for 480i and 576i resolution videos. The composite video only has 1 channel while the S-video version has two and the component video has 3 channels. Since video quality increases with more channels then expect video quality to be average with the Raspberry Pi.

Audio docket: the hardware in the Raspberry Pi where audio jacks are connected to allow the device to process audio materials.

Power supply: The hardware used for power supply storage. This is the part of the Raspberry Pi where the battery charger is plugged into to enable the device to store power for the device to function.

USB port: Universal Serial Bus is the connector that will allow other electronic devices such as digital cameras, computers, MP3 players, chargers to communicate and connect with the Raspberry Pi.

General Purpose Input and Output (GPIO): Integrated circuit consisting of generic pins usually assembled in a row on a circuit board. The GPIO pins add an element of versatility on a circuit board. It gives the programmer the ability to send control signals to the board making it perform actions based on programming.

Check the circuit board of any commercial computer and it is likely the same hardware can be found in it. This means the Raspberry Pi is capable of working in much the same way as any personal computer available in every computer sales shop.

Other Things You Need

The Raspberry Pi Foundation made the device available to anyone interested in the technology. It can be purchased at a cost of $25 to $35 depending on the model preferred. A no excuses price tag to getting people interested in coding by making it a very inexpensive investment.

The next step after purchasing a unit of Raspberry Pi is to gather external hardware to enable assembly of a personal computer. After all a simple circuit board does not a computer make and the Raspberry Pi is not a computer until it is connected to these basic external hardware – monitor, keyboard, and mouse, to name a few.

But take heart, all of the hardware required to create a computer are very likely available in every household. It will not require a mojr investment to build one using the Raspberry Pie.

What external hardware does the Raspberry Pi require? Here is a list - monitor or TV set, mouse, USB hub, keyboard, SD hub, hard drives, power source, speakers, cables, and case. It is a simple enough list. In fact, if people looked around their house or office it is likely they will find all these components there.

More details on the basic elements to get a fully functional Raspberry Pi powered computer running in minutes:

1. Monitor

    Pi does not have its own monitor; therefore, to control the interface, you must have a monitor connected to Pi via a high definition multimedia interface (HDMI) connection. Most monitors have a DVI port instead of an HDMI port. If this is the case, you have to buy a DVI-to-HDMI converter cable.

2. TV

    The Pi interface is not only compatible with computer monitors for visual display. High-definition TV usually has HDMI ports that can be connected to Pi using an HDMI cord to give you a crisper display. Old televisions can be used but you will have to use an HDMI-to-RCA cable. In this case, text output does not have a sharp definition making it difficult for any user to read.

3. USB Hub

    The board only allows for two USB ports, which you can use to plug in your keyboard and mouse leaving none for other hardware installation. To address this issue, you can look for a USB Hub to allow several other devices to be connected to your Pi. Make sure that the USB Hub generates power from a source independent of the Raspberry Pi to provide the Pi with added power and to prevent difficult circumstances.

4. USB-powered keyboard and mouse

    Raspberry Pi only supports USB keyboard and mouse. Often, Pi misbehaves because keyboard input devices derives lot of power from the Pi so make sure that your keyboards does not have too many unnecessary features.

5. SD Card

    This should play as its main storage device considering that Pi does not have a built-in hard disk. Although you probably have SD cards for your DSLR, you will probably need one that has a higher memory capacity. This book recommends that you purchase a card that has a memory equal or more than 8 GB. Even though you will find out that this may not be as powerful as that of modern PC, you can still use an external hard drive for storage.

6. Flash drives

Since you only have little memory to work with, you can use Flash drives or memory sticks to give you additional space. They now come in high-capacity and incredibly-cheap releases. This also enables you to transfer files easily between Pi and your PC.

7. External hard drives

If you still feel limited by the memory offered by flash drives, you may also want to invest in an external hard drive where you can install your music and video collections. External hard drive also uses USB port but it may use a lot of power. Even more reason to purchases a powered USB hub especially if your external hard drive does not have an independent power source and depending on other utilities to supply it with power.

8. Speakers

Raspberry Pi has an audio out port that is compatible with most PC speakers and headphones. If you are using TV connected via an HDMI connect, you'll find out that you don't need separate speakers because the audio is sent side-by-side with the video to the TV. This does not work however with DVI.

9. Power Source

Pi uses a Micro USB connector as a connecter to a power source. This is similar to most tablet and mobile phone chargers. It is possible that a lot of chargers do not have enough resistance to completely power up your Pi. If this is the case, make sure to buy charger from the same distributor as your Pi.

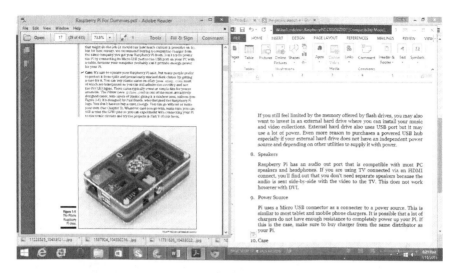

Figure 2: Pi Case

10. Case

    In order to protect Pi from any damage, you have to invest in a case that is sturdy enough to withstand force. Plastic cases can be bought online while others can be bought in the same store where you bought Pi.

11. Cables

    For the Raspberry Pi project, you will need a lot of cables to connect devices to Pi. For connecting monitors, you will need an HDMI cable. If you are using a DVI monitor, you will need an HDMI-DVI converter. For audio, you will need an audio cable. For internet connection, you will need an Ethernet cable.

Raspberry Pi is engineered to use several resources to make projects that you never thought possible. For more information about which devices are supported, you can check: http://elinux.org/RPi_VerifiedPeripherals

Raspberry Pi Foundation Designed Accessories

If you have a bigger budget and want to invest on a Raspberry Pi computer system with a more uniform and streamlined assembly. There is also an option to purchase Raspberry Pi designed accessories. The Raspberry Pi people have designed and developed all the necessary accessories required to create a Raspberry Pi powered computer. The list of accessories include the Pi Case, power supply, camera, USB hub, power supply, and even a 7" touch screen monitor.

Here is more info on the Raspberry Pi Foundation designed accessories:

The Pi Case

The wait is over. RaspBerry Pi has launched the official design of the Pi Case. It has high quality ABS construction with removable side panels and lid. This provides quick access to its GPIO, display connectors and camera.

The Pi Power Supply

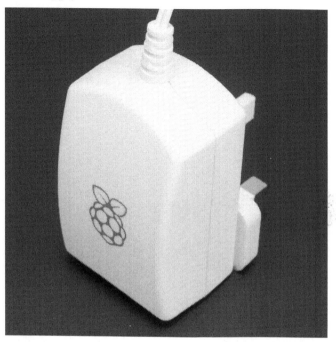

The RaspBerry Pi recommended universal USB micro power supply, it has gone through and passed a variety of tests. It will ensure you have a steady source of power for great performance. Its features include 1.5m lead, 50,000 hours MTBF, short circuit over current & over voltage protection, and 1 year warranty.

The Pi camera module

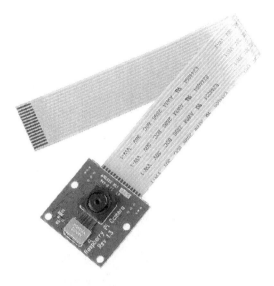

The Pi camera is a 5 megapixel fixed focus camera with 1080p30, 720p60 and VGA90 video and stills mode. It can be attached to the RaspBerry Pi CSI port using a 15cm ribbon cable. The camera takes high-definition video and stills photos. Its very user friendly for beginners and also has enough advanced features to interest the more sophisticated programmer. It's been used for some great time-lapse and slow motion video projects as well as effects driven work using the camera's create effects.

The camera is compatible with all Raspberry Pi versions.

The Pi Noir Camera

For more specialized video projects such as monitoring animal activity in the dark of night or installing security cameras to check for night activities, the Pi Noir camera is the solution.

As the name implies, Pi Noir means **No I**infrared. It is a camera that is optimized to shoot high-definition videos and photos in the dark. Expect daylight photos and videos captured with this camera to look a bit off since it does not have the usual camera features for a daylight shoot. But it delivers great picture for projects shot in the dark.

The Pi Noir Camera is compatible with all Raspberry Pi 1 & 2 models.

The Pi USB

This is the recommended universal USB port for all Raspberry Pi devices as it works so much better compared to non-Pi USBs. It is designed to be more compatible and therefore generate more optimized performance when used alongside its sister devices.

Its features are 802.11b/g/n, dimensions 30x16x8mm including USB plug, BCM43143 chipset, built-in support in NOOBS and RaspBian, and 150Mbps maximum throughput.

The Sense Hat

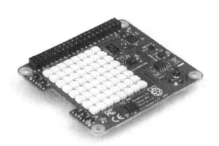

With an 8x8 RGB LED matrix and 5 button joystick containing the following sensors gyroscope, accelerometer, magnetometer, temperature, humidity, and barometric pressure. It is an add-on board for the Raspberry Pi models. In 2015 the Sense Hat had the distinction of joining the team of astronauts in the International Space Station.

The Pi Touch Screen Monitor

This is a 7" touchscreen monitor with 800x480 display connectors via an adapter board capable of handling power and signal conversion. Not necessary to plug in a keyboard and mouse with the Pi touch screen monitors. Transform any Raspberry Pi device into a touch screen tablet, standalone device or infotainment system with this.

CHAPTER 3

The Operating System

Before you start any Raspberry Pi project, you still have to install an operating system (OS) in for it to work. The OS allows you to access basic resources to simulate the functions of a computer and perform several activities including running programs and managing files. Computer applications, like web browsers and word processors, are assisted by the OS who acts as a liaison between the applications and the hardware. This set-up isn't exclusive to Raspberry Pi since laptops or computers are run by either Windows or Mac for an OS, phones and tables are either iOS or Android.

In this chapter, this book will teach you how to introduce an OS using an SD card. The process won't take long but it can be complicated as you will have to use commands and software that may not be familiar.

Some SD cards can be bought with a preloaded OS; however, mastering the skill of loading an OS to an SD card will help you when you want to try other versions of Linux.

Distribution

Several Linux distribution packets can be used for Raspberry Pi. If you want a list of accredited distributors, you can go to: www.raspberrypi.org/downloads. It shouldn't come as a surprise that there are other distributors with a lot of them in different stages of development and accessibility. For an appended list, you can go to www.elinux.org/RPi_Distributions.

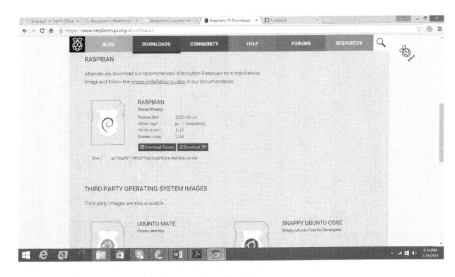

Figure 3: Accredited Linux Distribution

For beginners, the most recommended distribution is the Raspbian Wheezy, one of the publicly-released versions of Debian, which has been developed by two Raspberry Pi aficionados Peter Green and Mike Thompson. Included in their distribution are several development applications, Midori web broser and LXDE graphical software. If you are only beginning to understand how Pi works then you have to use Wheezy.

Another distribution is named the Arch Linux ARM, which is favored by experienced users because it gives them better control of the software they install. It is several times more complicated than Wheezy but it is more customizable. Once you get a better understanding of how Raspberry Pi works, you will want to switch to ARM because of the amount of control it can give you.

When loading an SD card with an OS, you must have an access to a computer. It doesn't matter whether you are using a Linux, Mac or Windows because the only thing you need is the ability to upload OS to SD cards. Here's how:

1. Download the distribution packet you want to use. The links are already mentioned above.

2. The downloaded file will give you a zip file so you will have to extract the file. Some downloaded content may contain multiple files but you'll only need the file with .img extension.

SD Card Flashing

Loading your SD card with Linux OS isn't the same as copying files to an external memory device. Linux distribution comes in a special extension format. This holds all the files that should be included in the SD card.

SD card flashing works to convert the .img file into a utility that will enable you to manage Raspberry Pu. Here comes the more complicated part because SD card flashing follows a different procedure depending on the type of OS in your system.

Windows

1. Go to: https://launchpad.net/win32-image-writer and download a copy of the latest release of Image Writer. Download the binary file, not the source file.

2. After downloading the zip file, extract the files to a separate folder.

3. Open the folder containing the extracted file.

4. Locate Win32DiskImager.exe and activate the executable file by downloading it. Allow program permissions.

5. Once a prompt opens, click on the folder icon next to the input space to open a file manager. Go to the folder that stores your Linux distribution, double click the image (.img) file.

6. On the device selection menu, choose the drive assigned to your SD card. Double-check this item because this step will be completely wiped out to make space for new data packets.

7. Click on the button that says, Write. The Image writer will then begin creating a usable Linux OS for your Raspberry Pi.

Mac

The difference between Windows and Mac is that the latter uses a script named RasPiWrite to simulate the process of SD card flashing.

1. Head to your Documents folder. Make a folder named SD Card Prep that contains a subfolder named RasPiWrite.

2. Pop open your Safari browser, go to https://github.com/exaviorn/RasPiWrite, and download the RasPiWrite zip file.

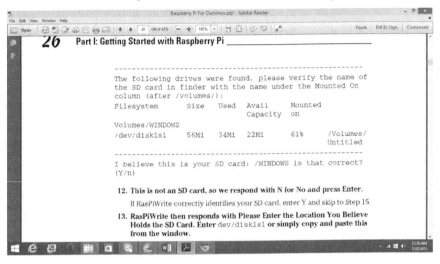

Figure 4: Warning Prompt

3. Once the download file is ready, open the subsequent folder produced by the download.

4. Copy the three files inside the folder to the RasPiWrite folder.

5. Copy the Linux distribution zip file you downloaded earlier into the RasPiWrite folder.

6. Press Command + Space, then type terminal to launch the terminal.

7. In the Finder window, locate the folder named SD Card Prep.

8. In the terminal, key-in cd succeeded by a space.

9. Drag the RasPiWrite folder until it finds itself in the terminal window.

10. Press Enter.

11. In the terminal, type ls (make sure that this is in lowercase) then press Enter to display a list of files contained within the folder.

12. Remove other external storage device leaving only the SD card plugged in your computer panel.

13. Run RasPiWrite by keying in the following: sudo puthon raspiwrite.py (Make sure that this is in lowercase.)

14. Since this entails overwriting a storage device, the system will prompt you for your password. Supply it.

15. After supplying your password, the above prompt will appear. If the system correctly identified your SD card, type Y but if it didn't, type N and check if there are other external storage device plugged into your computer.

16. The terminal will ask you for the location of the SD Card. Simply enter this: dev/disk1s1

17. Confirm that the location you supplied is correct by entering: Y and then pressing Enter.

18. It will then prompt you if you want to download a Linux distribution packet but since were done with this step, you can simply enter N and continue by pressing Enter.

    This step may take a while but once it is done, double-check if you have chosen the correct SD card.

19. Key in Accept in the terminal, then press Enter.

Linux

If you have access to a Linux-powered computer terminal, use this to flash your SD card as you won't have to download any additional software.

1. Remove other external device plugged in your computer.
2. Insert the SD card where you'll install the Linux distribution packet. It is important that this SD card is either empty or that you no longer need the contents stored in it.
3. Access Linux's terminal window.
4. Key in: sudo fdisk –l (note that's not 1, that's the letter "l") This will list down the names of the disks (internal and external).
5. Scour the list and locate the disk assigned to your SD card.
6. Key in cd to reach the directory that you used to store the image file of the distribution packet.
7. Key in ls *.img to locate the label of the image file on screen.
8. Manually write the image address to the SD card by keying in the following: sudo dd if=distribution.img of=/dev/sdX bs=2M

CHAPTER 4

Integrating Raspberry Pi 2

It's easy to get intimidated by the seemingly complicated work of art that is the Raspberry Pi but there's no reason to be. In fact, once your down with the steps enumerated in the previous chapter, connecting Raspberry Pi with compatible devices is easy. All you have to do is tweak its default configurations. Once you get the hang of it, you'll find that you are now capable of making a myriad of applications and projects.

Before you start tinkering with your device and integrating Raspberry Pi 2 to various devices, make sure that the side, where the Raspberry logo is printed, is facing up.

Here's what the upside should look like:

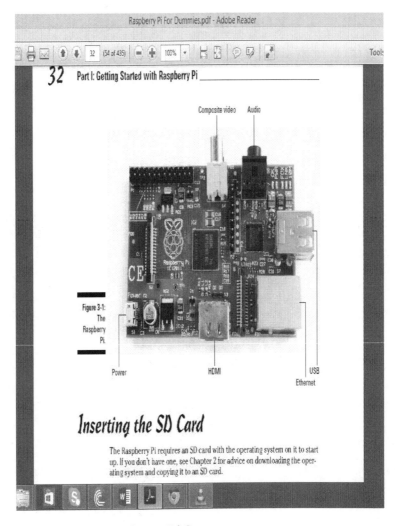

Figure 5: Parts of Raspberry Pi 2

SD Card Integration

Raspberry does not have an operating system installed in it so you have to integrate the SD card containing the operating system into your Raspberry Pi.

Flip the Raspberry Pi so that you are looking at the bottom portion. There, you'll find a plastic slot fixture where you have to insert the SD card. Don't force it in and be gentle when you slide the card enough to make sure that the system is connected well. Don't be surprised if it most of the SD card sticks out on the side of the board such that it you can still see a part of it even after flipping the Pi back to its upright position. Consider Figure 6.

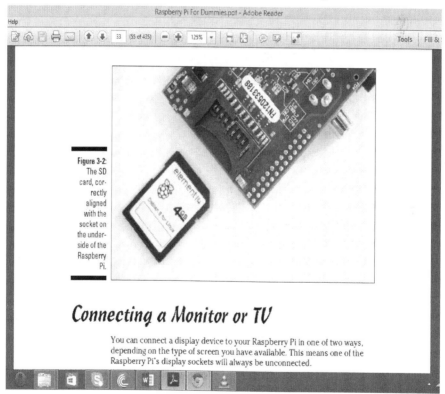

Figure 6: The Raspberry Flipside

Turning-on Raspberry Pi

Connecting the power on is the last thing that you do but for the sake of academic discussion, this chapter will tell you how to turn on your Raspberry Pi so you won't accidentally switch it on while connecting other units to your Pi board.

On the bottom left corner of Pi, you'll find a USB Power port, which you can use to insert a high Mhz charger to connect Pi to a power source. Raspberry Pi does not have an on and off switch so when you plugged in the Micro USB cord to a power source, the board lights up and starts working. The only way to turn it off is by disconnecting it from the power source.

Raspberry tells you that it's plugged in and running when the screen shows a brief color rainbow before loading up the Linux OS on the SD card. When it displays unintelligible words, do not panic because it is really meant to display these texts initially.

Connecting Visual Displays

Depending on the monitor that you have, there are two means for you to connect Raspberry into an output display device.

1. Connecting HDMI or DVI

    Raspberry operates in limited port possibilities, which fortunately includes a provision for HDMI connection that can be connected to a monitor display that can be either an HDMI port or a DVI port. If it is the latter, you must have an HDMI-to-DVI converter as in Figure 7.

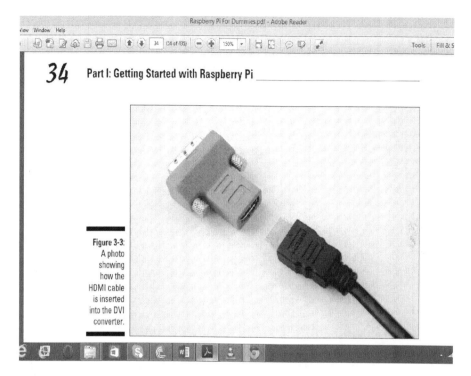

Figure 7: HDMI-to-DVI converter

2. Composite Video

   Television with HDMI ports will yield optimal display, however you can also use the silver and yellow port on the upper edge of the Pi designated for composite video. (Check Figure 5) If you have an RCA cable, connect it to the Video In port on the television. Switch to video mode to view the display from Raspberry Pi. Wait until you have powered up Raspberry Pi before checking the display.

Connecting USB Hub

As indicated in Figure 5, the Raspberry's USB port is located at the right edge of the Pi board. As there are, at most, two USB ports available, you can use a USB Hub to allow more devices to be plugged into Raspberry. It is important, however, to make sure that these devices have their own power sources. In Figure 8, you will find an example of a USB hub. The two USB ports in front allow you to use another extra port. The circular port in front can be used for charging the Hub preventing the usurpation of power from Raspberry.

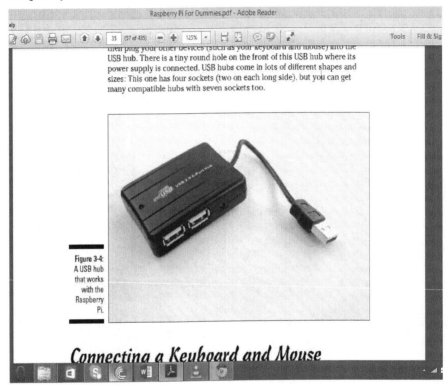

Figure 8: USB Hub

Connecting Audio

If you are using a display output that supports HDMI connection, then the sound is automatically routed to the television stereo hence there's no need for an extra audio cable. If you need a more discreet way to hear the sound output from Raspberry Pi, there's an audio socket where you can insert the round audio jack connected to a headphone or a mobile stereo.

If you are using mobile speakers, they should have their own power source.

Figure 9: Speaker Cables

Connecting Keyboard and Mouse

Most of the newest releases of mouse and keyboards are connected through the USB ports. It is better that both the mouse and they keyboard should be connected to the Pi via a USB hub to prevent both devices from draining power from the Pi board.

Connecting to the Internet Router

Unlike the first release where only Model B has Ethernet socket, Raspberry Pi 2 allows LAN connection via Ethernet port on the right edge portion of the

Pi board as shown to you in Figure 5. Use this port to connect your Pi board to the router using an Ethernet cord.

Dynamic Host Configuration Protocol (DHCP)-supported router automatically connects Raspberry Pi without having to configure anything.

CHAPTER 5

Linux

The best way for beginners to get the hang of Raspberry Pi is to use a visual desktop environment. One of the more popular interfaces that allow visual control over the Pi environment is called Lightweight X1 Desktop Environment (LXDE), which is an integral fragment of the Raspbian Wheezy distribution.

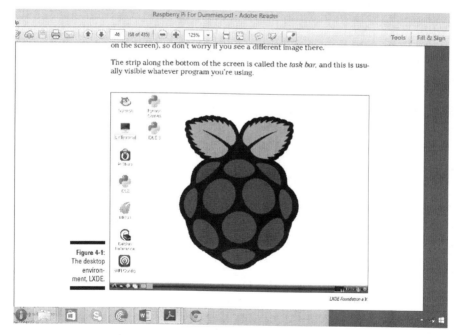

Figure 10: LXDE

The LXDE is specifically designed for beginners to use efficiently the memory and processors that makes Raspberry Pi work effectively despite its obvious limitation compared to other modern computers. LXDE works similarly as Windows and Mac OS does. They both let you use icons to manage and operate various files and programs.

Desktop Environment

After turning on Raspberry Pi, the monitor or visual output will first bring you into a command line prompt, which allows you to control the whole system by coding instructions in it. Depending on how you set up your computer, you may be required to input a password first especially if you specifically set it up to do just that for security.

To get to the desktop interface, you should type the command: startx in the command line prompt. The screen will go blank for a few second but it takes a few moments before the desktop environment starts to load.

The LXDE environment should ideally look like Figure 10 however, the desktop background can be changed. The task bar at the bottom should display which program you are currently using.

When using LXDE, you'll notice the different icons on the left side of the screen filled with various icons representing the programs and applications that you can use for your projects. These programs usually include the following:

1. Scratch: This programming language can be used to create animations and games.

2. Pi Store: This program gives you access to downloadable utilities that are compatible with your Pi board. Activate the icon on the left. It will prompt you to ask permission to read and write files within the memory space available in the SD card

3. LXTerminal: Whenever you need to key in coded instructions through a command line without shifting out of the desktop interface, you may do so using this key program.

4. Midori: This is the web browser exclusively made for Linux.

5. IDLE and IDLE 3: These program utilities are specifically made for Python Programming.

6. Wi-Fi Configuration: This allows you to build a wireless Internet connection when a Wi-Fi dongle is plugged in one of the unused USB ports.

7. Python Games: These games are made through Python programming for your gaming pleasure.

8. Debian Reference: This gives you help resources that you would find helpful as you play around the Raspbian Wheezy distribution.

When using one of the programs, double-click the icon that represents it. This is similar with Mac and Windows OS.

Task Manager

Since your Raspberry is working on a simpler system, it should not come as a surprise if it stops responding at times. This means that the system is running a lot of more complicated instructions that it can't process more input. The bottom right portion of the task bar houses an indicator on how much of the Central Processing Unit is being used in performance of the current functions. If it shows a bar chart then the rightmost bar is the indicator of the latest processor performance. If it has peaked, then it is more likely for the program or unit to crash. Always check the CPU Usage Monitor so that you'll know how to control your usage so that the display won't freeze on you.

The task manager is responsible to list down all the running programs to give you a brief idea of how the programs are performing the tasks that are assigned to them. Similar to most OS, the task manager can be accessed by holding down Ctrl + Alt while pressing Delete. The prompt should look like this:

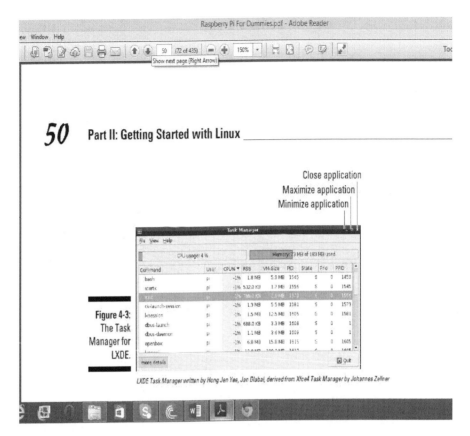

Figure 11: Task Manager

Similar to how you use a typical desktop interface in Mac or Windows, you simply right-click the item within the list of tasks enumerated by the task manager and choose Quit to terminate the program. This shuts down the program without incurring unprecedented effects to the system.

File Manager

Although the command line allows you to manage your files, the desktop environment is more visually-appealing, making it easier for you to use LXDE. File managers are tasked to handle the management (browsing, renaming, cutting, copying and deleting) this various folder, files and utilities on external storage devices or within the Pi system.

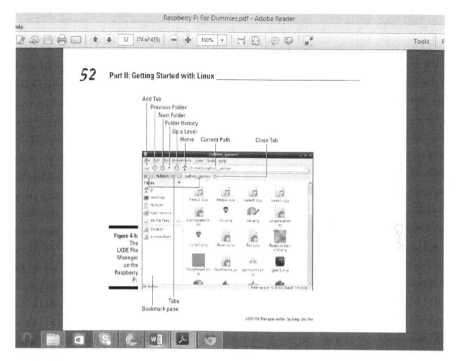

Figure 12: LXDE File Manager

Open the File Management system by clicking the bottom left button on the screen or locating it through Programs and within the category of System Tools. If you are used to using Linux, this distribution calls directories as folders just like Windows do. A folder houses several files that can be alike or different from each other. Just like directories, which can store subdirectories, folders can store folders too.

File Managers stores various kinds of folders including these four special main folders:

1. Desktop Folder: This holds the various files and programs that can be seen on the desktop usually for easy-access.

2. Pi Folder: This is where programmers tend to store important and crucial files and programs. This is the only folder that enables you to manage various documents and files as a typical user (as opposed

to administrators).

3. Rubbish Bin. This stores files temporarily until you decide to empty the Rubbish bin and erase all items permanently. It allows you to restore previously deleted file but only those that were yet to be deleted permanently.

4. Applications folder. This provides you access to the most important programs in the system.

There is no need to touch on how to transfer files from one folder to another nor copy files to a different storage device, nor to select multiple files, nor create blank files and new folder as these functions are similar to how you manage files and folders in your main computer.

External Storage Device

When accessing a desktop environment, you can use an additional storage device such as flash drives and external hard drives. Figure 13 shows you the dialogue box that appears once the system recognizes the storage devices. Using the File Manager in the previous section, you can open the contents of the storage device. Also, you ca modify its contents and add files to it:

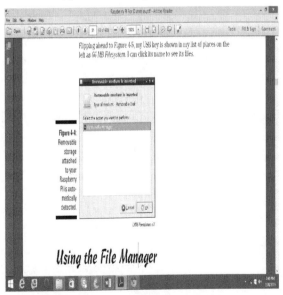

Figure 13: Removable Storage

Linux Prompt

Remember in the earlier part of this chapter, when you turn on your Raspberry Pi device, the screen will light up and the display will show this:

pi@easpberrypi ~ $

It may look like it's simply gibberish and unnecessary, but this display actually tells you more information than you think it does.

pi	This is supposed to show the log ID of the current user. In other words, raspberry allows various log ins of multiple users but at different times.
raspberrypi	This is the machine's name or how other devices – like your main computer – recognizes raspberry pi. The system allows you to change the device name.
~	Since you can manage your file using a command line prompt, this shows what the current directory is. For instance, the horizontal wiggly dash line (~) signifies the home directory.
$	This means that you are using the system as an ordinary user and not as an administrator; otherwise, the prompt would show a pound (#) sign instead of a dollar sign.

Files and Directories

Browse through the various files and directories within the system. Since you are currently restricted (as an ordinary user), you are barred from damaging any of the important files within the system.

When using the command line, access the list of files and directories by keying in:

ls

If you are currently at the home directory, you will be shown the folders and files within the home directory. To differentiate your input codes from the computer output, this book will highlight computer output.

pi@raspberrypi ~ $ ls

Desktop

Files

Python_codes

python_apps

Linux, unlike other OS, is case-sensitive. In other words, Linux sees Ls, lS, LS and ls differently. If you entered a command with wrong capital letters, the computer won't churn out the desire output. The same goes for files and directories.

Assuming Desktop is a directory, this means it may be containing various files within its system. In order to view the file listings inside the directory, you can change the current directory (home directory) to the Desktop directory by:

pi@raspberrypi ~ $ cd Desktop

The prompt will then show you the new current directory:

pi@raspberrypi Desktop $

Likewise, you can view the files within the directory by using the command ls as in this command line:

pi@raspberrypi ~/Desktop $ ls

File Types

If you need to gather information regarding several files, you can use the file command to display the file specification of certain files. Consider this example:

pi@raspberrypi ~/Desktop $ file remember.gif starwars.mp4 werq.py

remember.gif   GIF image data, 100 x 100, 16-bit/ color RGBA, interlaced

starwars.mp4:   mp4 data, WAV audop, Apple PCM. 16 bit, 880000 Hz

werq.py   Python script. ASCII executable text

This command is very handy especially if you need particular information about a file for proper file management. If you have been using a computer for a lot of time already, you can guess what the file is from just their extensions. For instance, .png, .gif and .jpg are usually image files.

The file command can also be used for directories. For instance:

pi@raspberrypi ~ $ file Files Desktop

Files: directory

Desktop: directory

This confirms that these objects are indeed directories.

Using the command cd, changes the current directory to one of the directories below it. If you need to go back to the parent directory or the directory above the current directory, you can use the command line cd .., like the example below:

pi@raspberrypi ~/Desktop $ cd..

pi@raspberrypi ~ $

When in the home directory (denoted by ~), you can try using the command line cd.. and see what will happen:

pi@raspberrypi ~/Desktop $ cd..

pi@raspberrypi ~ $ cd..

pi@raspberrypi /home $ cd..

pi@raspberrypi / $

When you turn on Raspberry Pi, you are automatically directed to the home directory but this doesn't mean that it sits atop the hierarchy of directory. In fact, if you try the command line cd .., you will reach another directory named /home. The backslash (/) in the directory name means that you have reach a drive, which is the second-highest directory in the hierarchy. If you use it one more time, you'll reach the root noted by a single backslash (/). This is the highest directory on the operating system.

To peek at the directories aside from /home, you can continue the above commands with this:

pi@raspberrypi / $ ls

This will list down the other drives stored within the root. The command prompt will display an output like this:

bin boot dev etc home lib lost+found media mnt opt proc root run sbin selinux srv sys tmp usv var

Aside from /home, these are the other drives within the root directory.

Root	Drive	Directory	Subdirectory
/	bin		
	boot		
	dev		
	etc		Files
	home	pi	Desktop
	lib		Python_codes
	lost+found		Python_apps
	media		
	mnt		
	opt		
	proc		
	root		
	run		
	sbin		
	selinux		
	srv		
	sys		
	tmp		
	usv		
	var		

The above table shows the hierarchy of tables, drives, directories and subdirectories.

Paths

When using file manager, you'll notice how the location of a file is denoted by the hierarchy of folders. This is how Linux keeps all files and directories organized to prevent overlapping of different files with the same file name. Now if you know the exact location of a file, you can supply the path name or the description of the location of the file within the hierarchy.

There are two kinds of paths. The first one is called relative path, which is giving the directions to the current directory whether you want to go up the hierarchy or below. The other one is called absolute path because it is the exact location of the file. This is similar to how street address works. Think of the country as the root; while the province or state is the drive; the city is the directory; the street is the subdirectory; and the house number is the name of the file.

For instance, if you are trying to get to the Desktop directory regardless of the current directory, you can simply key-in:

cd /home /Desktop

Remember that Linux is case-sensitive so make sure that you practice appropriate capitalization.

If you want to go ahead to werq.py, which is installed within the Desktop folder, you can key-in:

cd /home /Desktop /werq.py

If you want to go to the root, you can just supply a slash:

cd /

Consequently, if you want to go to the home directory:

cd ~

CHAPTER 6

Raspberry Management

After getting used to the processes that come with Raspberry Pi, you'll find installing new software, discovering various applications, and coding which are all integral to managing Pi boards properly less like a daunting task and more like a fun challenge.

When working with various Pi boards project, you may require packages for your codes and hardware devices to work properly. Fortunately, Raspberry Pi has a package manager called apt, which takes care of all the version dependencies, utility downloads and software installation.

Unlike other Linux distributions, the Raspbian Wheezy does not have a root account or an administrator account because it supposedly poses a threat to sensitive files and data when people forget to log out of the root account. There are commands that require approval of root account but in its absence, you can bypass said authority by including the word sudo before the command line. This command enables installation of programs that you might need for your Raspberry project.

Installing and Managing Software

Before you can install software in Raspberry Pi, you are required to update the list of packages within the package manager. Update the package repository by entering the following command:

sudo apt-get update

For this to work, you must have a stable internet connection. This can take time so make sure you have ample to wait for the updates to finish.

If you already have prior knowledge about the package name of the program you want to install, this command will download the program package from the net, including all the dependencies (packets of files needed for a package to work):

sudo apt-get install <name of package>

When you are locating a cache, you should use the command apt-cache instead of apt-get, which is used for installing software.

Linux allows user to run programs by just typing their names. Other applications require you to access them in your desktop interface.

One of the good things about package manager is that it makes sure that the software programs installed within the system are updated. You can issue a coded instruction so that the package manager updates all the software installed in the Pi board by keying in the following command:

sudo apt-get upgrade

If you want to make sure that the caches are upgraded to make sure that apt will be able to install the latest improvements introduced by the OEM to the installed programs, combine command lines to form a single-line coded instructions similar to this:

sudo apt-get update && sudo apt-get upgrade

The double ampersand (&&) instructs Linux to perform the second command line only if the first one is carried out successfully. In other words, if the cache failed to update then it won't update the software to their latest versions.

Package managers are also responsible for removing software from your system. Perform program deletion by keying in:

sudo apt-get remove <name of program to be deleted>

Deleting a program may leave few signs that it was there before, including file settings and saved games. If you want to remove the program and all its dependency files completely without leaving a trace, use this command line instead:

sudo apt-get purge <name of program to be deleted>

If you want to get a complete list of the programs installed within your system, you can use this command line:

dpkg –list

Since this does not involve installation or deletion of software, you are not required to procure root authority to run therefore, there's no need to bypass the root authority system by including sudo.

If you can find a specific package, you can run a search if they are installed by using this command line:

dpkg –status <name of package>

User Accounts

Even if the resources available for Raspberry Pi can be limited, you can share your projects with friends and families by creating separate user accounts for each one. Linux has a robust secure permission system that makes sure that each user won't be able to modify each other's files "accidentally".

Linux allows you to create groups and allow permission setting unique to each group. Even before modifying the capabilities of an account, you must be able to check out the group where you assigned a user. For instance, if you want to determine which group a certain member named Warren belongs to, you can use this code:

pi@raspberrypi ~ $ groups Warren

Warren : pi adm cdrom audio video plugdev games sudo dialout users input
    netdev

This means that user Warren has control over the pi directory, CD Rom, audio, video or almost every function enumerated in the output line of the above command line.

When creating a new user account, you have to make sure that this account belongs to a group with predefined function. Follow this format when adding a new user account:

sudo useradd –m –G <group name> <username>

In the above command line the –m is used to create a new home directory for the user, while the command -G makes sure that the member will belong to an already predefined group.

CHAPTER 7

Using Raspberry

In the first five chapters, this book has discussed extensively how to get Raspberry Pi 2 to work from what it needs in the inside (Operating System software, data packages, programming know-how) and outside (external device connections and Pi hardware), now is the time to put what you've learned so far to the various applications that are often linked to Raspberry.

Libre Office

Probably the most pronounced function of computers in the whole world is processing word documents. Raspberry allows you to do just that through LibreOffice, the successor of the famed OpenOffice. If you were familiar with the Microsoft Office then you wouldn't find it hard to use LibreOffice.

To install LibreOffice, initiate these command lines:

sudo apt-get update && sudo apt-get install libreoffice

If you have successfully installed Libre Office within the system, then you can use the five main program it offers: LibreOffice Writer (word-processing), LibreOffice Impress (presentations) LibreOffice Draw (graphic design), LibreOffice Calc (spreadsheers) and Libre Office Base (database system.

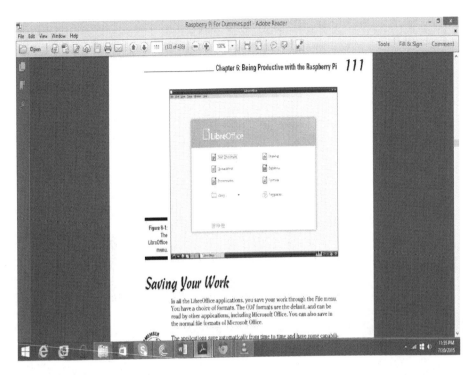

Figure 14: LibreOffice

When using LibreOffice, it is important to save your work to prevent the loss of valuable information.

GIMP Photo Editing

In the era where almost everything can be published in the internet through Facebook and other social networking sites, photo editing has evolved to be a very lucrative application that saw the likes of Adobe Photoshop stood out. Fortunately, Raspberry Pi has its own version in what we call GIMP.

In order to install GIMP, type the following:

sudo apt-get install gimp

At first, GIMP, could look too intimidating but once you get the hang of it,

RASPBERRY PI 2

you will be able to edit photos with ease.

Figure 15: GIMP

CHAPTER 8

Raspberry Pi: More than a personal computer

Like any simple computer, the Raspberry Pi is composed of the following basic computer circuit board features - an SOC which contains a central processing unit, graphics processing unit, and RAM. These make up the heart of the circuit board, the part that gives life to Raspberry Pi programming.

It also has an SD card slot which provides important storage space necessary in every computing machine. And of course the most important hardware on the Raspberry Pi, the GPIO or the general purpose input and output pins. A Raspberry Pi feature which gives it versatility. It allows the circuit board to be more than just a personal computer but so many other sorts of automated devices the inventor and programmer wish it to be.

The world should be celebrating the ability of this very simple piece of technology to let ordinary people to explore the realm of coding and computing. It puts them in a comfortable position within the world of computer technology by giving them the knowledge to take a device such as the Raspberry Pi and create something great with it.

Now we know all about the Raspberry Pi, and if inclined are equipped with the basic technical knowledge to create a personal computer using the Raspberry Pi circuit board.

After the in depth discussion of everything technical about the Raspberry Pi it should already be clear the chip has immense potential for computing. Knowledge that should make any person call Raspberry Foundation and ask to purchase a unit. Hopefully, sooner than later, people will be hard at work testing the Raspberry Pi for its many possibilities for developing all sorts of technology devices.

We understand now that several models of the Raspberry Pi are available in the market. And depending on the purpose it will be used for or the device that will be built with it we know there is a wide range of Raspberry Pi models to choose from to make said devices.

Again a reminder, the Raspberry Pi is not merely a circuit board limited to creating a personal computer. It is more than a personal computer. At best, it is a computing machine capable of automating anything provided it is given the right programming.

For example, if an inventor wants to create a remote controlled or computer activated gate for a house the Raspberry Pi can be programmed to create a computer automated gate.

If someone from the creative industry wants to build a camera that can capture time-lapse videos at the press of a button for creative projects, it is possible to do so with a properly programmed Raspberry Pi.

If a music producer wants to develop an MP3 player that can work according to specific requirements of a music project, the circuitry of the Raspberry Pi is capable of creating said MP3 player.

Or if a British astronaut on a mission in the International Space Station wants to listen to music in space, he can issue a challenge to young students in the UK to use Raspberry Pi technology to develop an ISS compatible MP3 player for him.

It has not been discussed in length enough what makes the Raspberry Pi so versatile. How is the Raspberry Pi able to create such a wide range of tech devices?

The answer of course is the GPIO or the General Purpose Input and Output pins located in the circuit board. The row of pins located on the edge of the Raspberry Pi. The GPIO because of its input/output capability gives the programmer the ability to tell the Raspberry Pi to act or perform actions that it programs it to do.

It's fascinating how 40 unassuming silver pins set in a row on a 3 x 2 inch piece of circuit board can allow programmers to do its bidding. The GPIO lets inventors create just about any gadget they want to create, not just personal computers.

The GPIO is the part of the Raspberry Pi that makes it an exciting piece of technology. With it the imagination is the only limit in terms of achieving the Raspberry Pi's complete potential.

Conclusion

Look at who's reached the last page!

For a quick recap, the Raspberry Pi is a technological marvel with the capacity to redefine the world with its ability to teach young kids, the next generation of people that will join the global workforce, to be adept at the language of computing. By breaking down the myth that computer science is an exclusive clique only the most intelligent people are allowed to join, the Raspberry Foundation intends to make coding & computing as easy to learn as the most basic general course children take in school.

This book is an introduction of the latest model of the Raspberry Pi series, the Raspberry Pi 2. With an upgraded SD RAM at 512MB on the Model B and 1GB on the Model B+. This new version of the Raspberry Pi allows it to be a faster and more efficient version to the earlier models. Upgrades that will give inventors and programmers more leverage to build more interesting tech programs and gadgets with.

The Raspberry Pi is a piece of technology that has the potential to automate any device it is connected to. With proper programming, it is a technology device that teaches kids and adults alike the limitless potential of computing and coding. It places in their hands the ability to create technology driven devices that can contribute to making a better life for people. Consequently, the Raspberry Pi has the ability to build a much better world through the power of automation.

Now that you managed to finish the book, I am confident that you will be able to apply your new knowledge into a worthwhile Raspberry Pi 2 project

This book only provided you with the basic information that you need to know to make a Raspberry Pi project but it is entirely up to your creativity to create something wonderful from something that would have been discarded by other people.

Would you do me a favor?

Finally, if you enjoyed this book, please take the time to share your thoughts and post a positive review on Amazon. It'd be greatly appreciated!

Thank you and good luck!

Made in the USA
San Bernardino, CA
27 February 2016